The bears make porridge
every day,
and always do it
a certain way.

AF700784

They cook the porridge
in a pan,
and then they follow
this simple plan . . .

Papa Bear is first
to count;
he always gets the
biggest amount.

Baby Bear always
gets the least —
really he'd like his
amount increased.

Mama's amount is
in between;
when she's finished,
the pan is clean.

But one day the plan
just wouldn't do —
someone else wanted
breakfast, too!

They needed to make
enough for four;
the bears had never
done that before!

But no one had to
go without —
clever Goldilocks
figured it out!